Setting: night, in the living room at Madge's house

Characters:

Madge: eight, the birthday girl

A.J.: Madge's friend

Liz: Madge's friend

Something strange *(out of sight)*

A.J.: Happy Birthday, Madge! Do you feel different now that you are eight?

Madge: No, not really!

Liz: The best thing about sleepovers is that you don't have to sleep! You can play games all night!

A.J.: Let's play a game now! How about hide-and-seek?

Madge: Sounds good. I'll be IT!

(Liz gives Madge a friendly nudge.)

Liz: Madge, get out of our sight so we can hide!

Madge: Are you afraid I might peek?

A.J.: Not at all!

Madge: I'll go into the other room.

Liz: Come right back when we call!

(Madge leaves. Liz wedges herself under a couch, and A.J. hides behind a chair.)

MADGE: Ready or not, here I come! I'm going to catch you!

(The lights go out. The room is pitch black. A.J. and Liz give startled gasps.)

LIZ: That's not funny, Madge! Switch on the light!

A.J.: You're giving us a fright!

MADGE: I didn't switch the light off. Something is not right!

(Suddenly there is a loud scratching noise. The girls all jump in fear.)

A.J.: What was that?

MADGE: It came from behind the door!

(The scratching noise gets louder.)

A.J.: Who or *what* is scratching like that?

LIZ: *(weakly)* A monster with an itch?

(There is a loud, moaning sound. Something whines at a high pitch.)

A.J.: What was that?

LIZ: I don't know, but it didn't sound friendly!

(The girls jump onto the couch and clutch each other.)

MADGE: Should we try to catch it?

A.J.: *You* can try. I won't budge!

Liz: Oh, no! The door is opening. Something frightful is there!

A.J.: It looks huge! It must weigh 100 pounds!

(There is the sound of panting, and running footsteps. The girls clutch each other more tightly.)

Madge: Here it comes! Get ready for a fight! ... Oh, no! It's got me!

(The lights go on. A dog is on Madge's lap, panting joyfully.)

A.J.: The lights are back on!

MADGE: *(laughing and petting the dog)* Our "monster" was Fudge!

LIZ: What a fright he gave us!

MADGE: What a night! Oh well, all's well that ends well!